THE BIRDS IN MY BACKYARD

A BEGINNER'S GUIDE TO CHICKEN BREEDS

By Alison Isaacs

Contents

The Origins of Chickens

Throughout history, humans and chickens have shared a unique and enduring relationship. Dating back thousands of years, chickens have played a significant role in human civilizations, serving not only as a valuable source of food but also as symbols of fertility, divinity, and companionship. From ancient rituals and religious practices to their practical use in agriculture, the bond between humans and chickens has transcended time and cultural boundaries.

In ancient civilizations such as Egypt, chickens were revered as sacred creatures associated with divinity. They were often sacrificed to honor gods and goddesses, and their feathers and eggs were used in religious ceremonies. The ancient Greeks believed that chickens had the power to predict the future through their behavior, leading to their use in divination practices.

As human settlements developed and agriculture became more prevalent, chickens became a vital source of sustenance. Domestication of chickens began around 6000 BC, and since then, they have provided humans with a reliable source of meat and eggs. Their ability to adapt to various environments, coupled with their high reproductive rate, made them an ideal livestock animal for many ancient civilizations.

Beyond their practical uses, chickens also became companions to humans. In ancient China, chickens were kept as pets for their beauty and unique personalities. They were often found in imperial gardens, where their colorful plumage and graceful movements brought joy and tranquility. In ancient Rome, chickens were kept in homes as a symbol of good luck and prosperity.

The relationship between humans and chickens has endured to the present day. In many cultures, chickens continue to be a staple in culinary traditions, and their meat and eggs are enjoyed worldwide. Furthermore, chickens play a crucial role in sustainable agriculture, providing natural pest control, fertilization, and a renewable source of high-quality protein.

The Red Junglefowl, scientifically known as Gallus gallus, is a remarkable bird species that serves as the wild ancestor of our beloved domesticated chickens. This vibrant and captivating bird is native to the dense jungles of Southeast Asia, where it roams freely, showcasing its striking plumage and displaying its impressive vocal talents.

With its signature fiery red and orange feathers, adorned with iridescent green and blue accents, the Red Junglefowl is a sight to behold. The males, known as roosters, boast long, flowing tail feathers and striking combs and wattles that add to their majestic appearance. In contrast, the females, known as hens, exhibit more earth-toned feathers, blending effortlessly into their natural surroundings.

Beyond their aesthetic appeal, Red Junglefowls play a significant role in the evolutionary history of chickens. These remarkable birds possess a remarkable ability to adapt to various environments, which has allowed them to survive and thrive for thousands of years. They are known for their agility, excellent flying capabilities, and their keen senses, which enable them to avoid predators and find food sources in their natural habitats.

The Red Junglefowl's vocalizations are both distinctive and enchanting. The roosters proudly announce their presence with their iconic crowing, a sound that has become synonymous with the break of dawn in rural areas around the world. This unique vocalization not only serves to assert dominance within their social hierarchy but also acts as a form of communication to attract mates and warn of potential dangers.

Due to their impressive physical traits and fascinating behaviors, Red Junglefowls have captivated humans for centuries. Their intriguing traits and behaviors have drawn the attention of early explorers, scientists, and eventually poultry enthusiasts, leading to their domestication and the birth of the countless chicken breeds we know today.

The domesticated chickens we now keep as companions, providers of eggs, and sources of meat owe their lineage to the Red Junglefowl. Through selective breeding, humans have shaped and diversified the genetic pool of chickens, resulting in a wide range of breeds with various physical characteristics and specialized traits.

While the Red Junglefowl continues to thrive in the wild, their domesticated counterparts have become an integral part of our lives, providing sustenance, companionship, and even serving as cultural and religious symbols across different societies. Recognizing the Red Junglefowl's importance as the ancestor of our beloved chickens highlights the remarkable relationship between humans and animals and serves as a reminder of the interconnectedness of all species on this diverse and beautiful planet we share.

The domestication process of chickens dates back thousands of years, and it is a fascinating journey that has shaped the modern chicken breeds we see today. Initially, chickens were domesticated from the wild red junglefowl (Gallus gallus) in Southeast Asia, particularly in present-day Thailand and Vietnam. This process likely began around 6,000 years ago, with humans gradually taming and selectively breeding these birds for various purposes.

Early on, chickens were primarily valued for their eggs and meat, as well as their feathers, which were used for clothing and decoration. As people began to settle into agricultural communities, chickens became a vital component of their self-sustaining societies. Over time, different regions developed specific breeds that were best suited to their needs and local environments.

One such early breed is the ancient Mediterranean breed known as the Leghorn. Originating in Italy, Leghorns were prized for their exceptional egg-laying abilities and hardiness. They quickly gained popularity across Europe and were later exported to the United States, where they played a crucial role in the development of the American poultry industry.

Another notable early breed is the Sussex chicken, which originated in England. Sussex chickens were highly versatile, excelling in both egg production and meat quality. They were known for their calm temperament and ability to adapt to various climates, making them a favorite among farmers for centuries.

As chicken domestication spread to different parts of the world, unique breeds emerged. For example, the Brahma chicken, hailing from the United States, became renowned for its impressive size and strength. With their feathered feet and gentle disposition, Brahmas were widely sought after for exhibition purposes.

The domestication of chickens and the development of early breeds laid the foundation for the diverse assortment of chicken breeds we have today. Whether it's the elegant plumage of ornamental breeds, the robust meat production of broiler breeds, or the superior egg-laying capabilities of specific layers, each breed serves a unique purpose and exhibits distinct characteristics.

From the ancient Leghorns to the majestic Brahmas, exploring the early chicken breeds not only provides insights into our agricultural history but also highlights the remarkable ingenuity of humans in selectively breeding and shaping the traits of these remarkable birds. As we continue to cherish and appreciate chickens in various aspects of our lives, let us remember the rich heritage and evolution that has brought us this diverse array of chicken breeds.

Since the scope of this book is only to acquaint the reader with some of the more popular breeds of chicken, one will certainly find the list which follows to be far from encyclopedic with unavoidable instances of overlapping between heritage, dual purpose, and ornamental breeds.

Chapter 1: The Heritage Breeds

Ameraucana

The Ameraucana breed, also known as the "Easter Egger," is a beloved breed among chicken enthusiasts. With its striking appearance and unique egg-laying abilities, the Ameraucana originated in the United States. The breed was developed in the 1970s, but its ancestors can be traced back to the Araucana chicken, a breed originating from Chile known for its blue eggs. The Ameraucana was selectively bred to improve its overall characteristics, including its temperament, hardiness, and distinct blue egg-laying ability.

The most visual feature of the Ameraucana is its vibrant plumage. It comes in a variety of colors, including black, blue, silver, and wheaten, with each color displaying beautiful feather patterns. The breed boasts a full beard of feathers, known as muffs, and a unique set of ear tufts, making it instantly recognizable.

Beyond its stunning appearance, the Ameraucana is highly regarded for its egg-laying capabilities. While many chicken breeds lay white or brown eggs, the Ameraucana surprises with its range of blue, green, and even pale pink eggs. This trait, inherited from its Araucana ancestors, adds an exciting and colorful twist to any egg collection.

In terms of temperament, the Ameraucana is known for being friendly, docile, and sociable. They generally get along well with other chickens and make for excellent additions to backyard flocks or family farms. Their calm demeanor and non-aggressive nature make them a popular choice for chicken enthusiasts of all levels of experience.

When it comes to care, the Ameraucana is relatively easy to maintain. They adapt well to various climates and are known for their resilience to harsh weather conditions. Proper housing, nutritious feed, and regular health checks are essential to ensure their well-being and egg production.

Cochin

The Cochin chicken, also known as the "Gentle Giants," is a remarkable breed originating in China. This majestic bird is cherished for its unique appearance and docile nature. With its feathered legs, voluminous plumage, and gentle demeanor, the Cochin chicken stands out among other poultry breeds.

Cochins come in a variety of stunning colors, including black, white, blue, buff, and partridge. Their large size and abundant feathers give them an impressive and regal appearance. Despite their size, Cochin chickens are known for their calm temperament, making them excellent additions to any backyard or farm.

These birds are not only aesthetically pleasing but also bring numerous practical benefits. Cochin hens are exceptional mothers, often exhibiting broody behavior and excelling in raising chicks. Additionally, their ample feathers provide excellent insulation, enabling them to withstand colder climates more effectively than other breeds.

Cochin chickens are versatile in their purpose as well. While primarily known as an ornamental breed, they also produce high-quality eggs. Though they may not be the most prolific layers, the eggs are large, brown, and deliciously rich. Whether you're seeking a charming addition to your flock or a delightful source of fresh eggs, the Cochin chicken can fulfill both roles with its unique combination of beauty and utility.

As a friendly and personable breed, Cochin chickens are often adored by children and adults alike. Their calm demeanor makes them easy to handle and interact with, making them an ideal choice for those seeking a more interactive poultry experience. These birds can become true companions, following their owners around the yard and even enjoying some gentle cuddles.

Dominique

The Dominique chicken, also known as the Dominicker or Pilgrim Fowl, is an iconic and historic breed that holds a special place in the hearts of poultry enthusiasts. Originating in the United States, specifically New England, the Dominique chicken is believed to be one of the oldest chicken breeds in America, with its roots dating back to the early 1700s.

This breed is known for its distinct black and white striped plumage, which has earned them the nickname "Little Barred Plymouth Rock." The Dominique chicken possesses a medium-sized body with a well-rounded shape and a confident, alert posture. The hens are known to be excellent layers, producing a respectable amount of large brown eggs throughout the year. The breed is also well-regarded for its hardiness, adaptability, and ability to forage, making it suitable for both rural and urban settings.

Beyond their practical qualities, Dominiques have endeared themselves to chicken keepers with their calm and friendly personalities. They are known to be docile, gentle, and easily handled, making them ideal for families and beginners alike. Additionally, their intelligence and curiosity make them a joy to observe in the backyard or free-range environment.

While the Dominique chicken faced near extinction in the early 20th century due to the rise of commercialized poultry farming, dedicated breeders and enthusiasts worked tirelessly to preserve and revive this remarkable breed. Today, the Dominique chicken continues to captivate individuals seeking a connection to America's agricultural past and those who appreciate a beautiful and dependable bird.

Hamburg

The Hamburg chicken, also known as the Hamburg fowl, originated in Germany during the 14th century, and quickly gained popularity for its striking appearance and exceptional egg-laying abilities. With its elegant plumage and charming personality, the Hamburg chicken has become a favorite among poultry keepers and chicken enthusiasts worldwide.

One of the defining features of the Hamburg chicken is its eye-catching plumage. This breed comes in a variety of colors, including silver-spangled, golden-spangled, black, and white. The plumage is glossy and lustrous, showcasing intricate patterns and vibrant hues that make the Hamburg chicken a true sight to behold. Whether it's the stunning contrast of black and white spangles or the elegant simplicity of the solid black or white feathers, each Hamburg chicken is a work of art in its own right.

Beyond their aesthetic appeal, Hamburg chickens are known for their friendly and sociable nature. They are curious and intelligent birds, often showing interest in their surroundings and displaying a strong sense of curiosity. Their amiable temperament makes them an excellent choice for backyard flocks, as they get along well with both humans and other chicken breeds. The Hamburg chicken's sociability lends itself to being a delightful addition to any poultry community, bringing joy and liveliness to the coop.

Aside from their beauty and friendly nature, Hamburg chickens are esteemed for their egg-laying capabilities. They are known for being prolific layers of small to medium-sized white eggs, consistently providing a steady supply of delicious and nutritious eggs throughout the year. This makes them a popular choice for those who prioritize both ornamental value and practicality in their flock.

Leghorn

The Leghorn chicken, also known as the Italian chicken, is a breed that has gained popularity worldwide for its impressive egg-laying abilities and distinctive appearance. Originating from the Italian port city of Livorno (formerly known as Leghorn), this breed quickly made its way to various corners of the globe, captivating poultry enthusiasts with its unique characteristics.

Leghorns are known for their rather petite size, with sleek, slender bodies and graceful movements. They possess a distinct triangular-shaped head, adorned with a prominent single comb and large, alert wattles. The breed showcases a wide range of stunning plumage colors, including white, brown, and black, with variations like buff, silver, and red too.

While the Leghorn's appearance is undoubtedly captivating, it is their exceptional egg-laying abilities that truly set them apart. These chickens are prolific layers, often surpassing other breeds in production. Leghorn hens can lay an impressive 280 to 320 extra-large, white eggs per year, making them highly valued for commercial egg production and small backyard flocks alike.

Moreover, Leghorns are known for their active and energetic nature. They are excellent foragers, always on the lookout for insects and other tasty treats in the yard. Their agility and ability to adapt well to different climates make them suitable for a wide range of environments, whether it be free ranging in the countryside or thriving in urban backyard coops.

Though Leghorns excel in egg production, they may not be as docile or affectionate as some other breeds. They tend to be more independent and can be a bit flighty, preferring to explore their surroundings rather than cuddle up with their human caretakers. However, with proper handling and socialization, Leghorns can become more accustomed to human interaction.

Plymouth Rock

The Plymouth Rock, also known as Barred Rock, is a beloved and iconic breed in the world of poultry farming. With its signature black-and-white barred plumage, the Plymouth Rock is easily recognizable. Its feathers create an attractive pattern that adds a touch of elegance to any flock. Not only is this breed visually appealing, but it also boasts several other desirable traits.

Plymouth Rocks are known for their docile and friendly personalities, making them excellent additions to any backyard flock or farm. Their calm demeanor makes them easy to handle and ideal for families with children. Additionally, they tend to be quite hardy and adaptable, thriving in various climates and environments.

One of the most significant advantages of the Plymouth Rock chicken is its exceptional egg-laying capabilities. Hens of this breed are prolific layers, producing large brown eggs consistently throughout the year. This makes them a popular choice among homesteaders and small-scale farmers who value a steady supply of fresh, nutritious eggs.

Furthermore, Plymouth Rocks are known for their moderate size and robust build. While they are not considered a heavyweight breed, they possess good meat qualities, making them suitable for both egg and meat production. Their excellent feed conversion and broad breast make them a versatile choice for those seeking a dual-purpose chicken.

Rhode Island Red

The Rhode Island Red breed originated in the United States, specifically Rhode Island, and has become one of the most popular backyard chickens worldwide. Known for its egg-laying abilities, hardiness, and gentle temperament, the Rhode Island Red is a popular choice for both novice and experienced chicken keepers alike.

With its vibrant mahogany-red feathers and sturdy build, the Rhode Island Red stands out in any flock. These chickens possess a single comb and bright red wattles, adding to their distinctive appearance. The roosters display an air of confidence, often strutting proudly with their chest held high, while the hens exhibit a graceful demeanor as they diligently go about their daily activities.

One of the most remarkable features of the Rhode Island Red is its exceptional egg production. Known as prolific layers, these hens consistently lay large, brown eggs, making them a valuable asset for those seeking a constant supply of fresh eggs. With an average of 200-300 eggs per year, the Rhode Island Red surpasses many other breeds in terms of productivity.

In addition to their egg-laying prowess, Rhode Island Reds are known for their hardiness and adaptability. They are well-suited to various climates and can withstand both hot summers and cold winters with ease. Their strong immune systems make them relatively resistant to common poultry diseases. These qualities, combined with their low maintenance requirements, make the Rhode Island Red a popular choice for backyard chicken enthusiasts of all levels of experience.

Beyond their practical attributes, Rhode Island Reds have endearing personalities. Despite being a larger breed, they are known for their friendly and docile nature, often becoming cherished companions to their owners. They are generally calm and easy to handle, making them suitable for families, children, and even urban settings where noise restrictions may apply.

Wyandotte

The Wyandotte breed, with its stunning beauty and charming personality, is a breed that stands out among the flock. Originating in New York state during the 19th century, the Wyandotte quickly gained popularity among poultry enthusiasts and farmers alike. Renowned for their unique laced feathers, which create a striking pattern, Wyandottes come in a variety of beautiful colors including silver, golden, blue, black, and buff. With their round, plump bodies, rose combs, and bright, inquisitive eyes, they exude an air of elegance and grace.

But it's not just their stunning appearance that makes the Wyandotte a favorite among chicken enthusiasts. These birds are known for their docile and friendly disposition, making them an ideal choice for backyard flocks, families, and even exhibition purposes. Wyandottes are excellent foragers and adapt well to various climates, making them a versatile breed for both rural and urban settings.

When it comes to egg production, Wyandottes certainly don't disappoint. Hens are reliable layers, producing medium to large-sized brown eggs that are both delicious and nutritious. Their hardiness and ability to adapt to different conditions make them excellent layers throughout the year, even during colder seasons.

Chapter 2: Exotic and Rare Breeds

Ayam Cemani

The Ayam Cemani, also known as the "Gothic chicken," or "Darth chicken," is a rare and captivating breed that hails from Indonesia. Renowned for its stunning all-black appearance, this unique chicken turns heads wherever it goes. Its feathers, beak, comb, wattles, feet, and even its internal organs are all black, thanks to a genetic condition called *fibromelanosis*.

Beyond its striking appearance, the Ayam Cemani is revered for its cultural and historical significance. Originating from the island of Java, it has a rich heritage dating back hundreds of years. In Indonesian folklore, the Ayam Cemani is believed to possess mystical powers, symbolizing good fortune and prosperity. It was even used in traditional rituals and ceremonies.

Aside from its cultural significance, the Ayam Cemani is also valued for its quality meat and eggs. The meat is said to be tender and flavorful, while the eggs are prized for their rich, dark yolks. Due to their rarity and high demand, Ayam Cemani chickens are often regarded as a luxury breed, sought after by collectors, poultry enthusiasts, and those who appreciate the extraordinary.

Raising Ayam Cemani chickens can be a rewarding endeavor, but it requires special care and attention. They are relatively small-sized birds with an active and energetic temperament. Adequate shelter, proper nutrition, and regular health checks are essential to ensure their well-being.

Brabanter

The Brabanter, a charming and eye-catching bird, hails from the region of Brabant in Belgium and has been admired for centuries for its striking features. The Brabanter is renowned for its impressive crest, which resembles a fluffy, feathery hat atop its head, making it instantly recognizable. With its elegant and upright posture, this breed exudes an air of regality and grace.

Beyond its aesthetic appeal, the Brabanter chicken is also highly regarded for being friendly, inquisitive, and sociable, making them an ideal addition to any backyard flock or family farm. Brabanters are not only wonderful companions but also reliable egg layers, producing a steady supply of large white eggs that are as delightful to look at as they are to savor.

Despite its smaller size, the Brabanter chicken is a hardy breed that adapts well to various climates. With its dense plumage and sturdy build, it can withstand colder temperatures, making it suitable for regions with harsh winters. Their ability to forage and thrive in free-range environments also makes them a favorite among enthusiasts of sustainable and organic practices.

For poultry enthusiasts seeking a breed that possesses both visual appeal and a gentle disposition, the Brabanter chicken is an excellent choice. Its rarity and unique attributes make it a fascinating addition to any backyard flock.

Faverolles

With its unique and charming appearance, The Faverolles is a beloved breed among poultry enthusiasts. Originating in France in the late 19th century, this gentle and friendly bird has captured the hearts of many with its abundant feathering. A profusion of soft, fluffy feathers covers its body with plumage which comes in a variety of colors, including salmon, black, blue, and cuckoo.

Beyond its appearance, the Faverolles chicken is known for its amiable temperament. These chickens are incredibly friendly and sociable, making them a perfect addition to any backyard flock or family setting. They enjoy human interaction and are often sought after as pets or exhibition birds. Children especially adore their calm nature, making them a great choice for families with young ones.

Furthermore, the Faverolles chicken is an excellent layer of medium to large-sized light brown eggs. Their consistent egg production, combined with their endearing personality, makes them a popular choice for both hobbyists and small-scale farmers. They tend to be quite hardy and adaptable, thriving in various climates, which adds to their appeal for poultry enthusiasts around the world.

Additionally, Faverolles chickens display a high level of broodiness, making them exceptional mothers. They are known to be attentive and caring, eagerly tending to their chicks with utmost dedication. Their nurturing instincts and ability to raise a healthy brood make them a treasured breed for those interested in breeding or hatching chicks.

Frizzle

The Frizzle chicken is a truly unique breed that stands out in a flock. Known for its curly feathers that give it a funky, frizzled look, this chicken breed is a delight to behold. Originally bred in Asia, Frizzles have since gained popularity worldwide for their quirky charm and friendly nature.

One of the most eye-catching features of the Frizzle chicken is its plumage. Instead of the typical smooth, straight feathers seen on most chickens, Frizzles possess feathers that curl outward, giving them a truly extraordinary appearance. This curly feathering extends from their neck all the way down to their fluffy, frizzled tail feathers. It's no wonder they capture attention wherever they go!

Beyond their striking appearance, Frizzle chickens are known for their easygoing and calm temperament. They are friendly, affectionate, and often enjoy human interaction, making them a wonderful addition to any backyard flock. Their docile nature also makes them a great choice for families with young children or for those seeking a peaceful and enjoyable chicken-keeping experience.

Frizzles come in a variety of colors, including black, white, blue, and even buff. Their unique feathering pattern, combined with their vibrant hues, makes them a favorite among chicken enthusiasts and hobbyists looking for something out of the ordinary. Whether you're a seasoned chicken keeper or a beginner, the Frizzle chicken is sure to add a touch of whimsy and charm to your flock.

In terms of care, Frizzles require the same basic needs as any other chicken breed. They thrive in a spacious and secure coop, and it's essential to provide them with a nutritious diet, fresh water, and proper veterinary care. Their curly feathers might require a bit more attention to prevent tangling or matting, but regular grooming and monitoring should keep them in good condition.

Polish

The Polish chicken, also known as the Paduan or Poland, is a unique breed of chicken. Originating in Poland, this ornamental bird is instantly recognizable for its distinct appearance: a striking crest of feathers on its head, resembling a regal crown.

The breed comes in various color variations, including white, black, silver, and golden. Its beautifully patterned plumage, combined with its majestic crest, makes the Polish chicken a favorite among chicken enthusiasts and a popular attraction in poultry shows.

Not only are Polish chickens a visual delight, but they also make excellent pets. Despite their elegant appearance, they are known to enjoy human interaction and are often sought after for their gentle temperament.

Though they excel in charm and personality, Polish chickens are not the most practical breed when it comes to egg-laying. They are considered average layers, producing a modest number of small to medium-sized white eggs. However, their unique appearance and delightful personalities more than make up for any lack in egg production.

In addition to their ornamental appeal, Polish chickens are known for their adaptability to different climates. They thrive in both cold and warm environments, making them suitable for a wide range of locations. Their hardy nature contributes to their popularity as a breed that can be enjoyed by chicken enthusiasts around the world.

Sebright

The Sebright has a strikingly unique appearance, standing out as one of the most captivating breeds in the poultry world. Named after its creator, Sir John Sebright, this breed originated in England during the early 19th century. Renowned for its compact size and exquisite plumage, the Sebright chicken has become a favorite among poultry enthusiasts and backyard chicken keepers alike.

What makes the Sebright chicken truly remarkable is its laced feathers, which give it an appearance reminiscent of delicate lacework. Available in two varieties, Silver and Gold, these chickens showcase a stunning display of intricate patterns and vibrant colors. Their feathers have a lustrous metallic sheen that catches the sunlight, making them a truly enchanting sight.

Beyond their natural beauty, Sebright chickens possess a friendly and sociable temperament, making them a joy to have in any flock. They are known to be docile and can often be seen happily foraging and scratching in the yard. Due to their small size, they require less space than other breeds, making them an ideal choice for urban or limited-space environments.

While the Sebright chicken is primarily admired for its ornamental qualities, their egg-laying capabilities should not be overlooked. Though they may not produce as many eggs as larger breeds, Sebright hens lay small but perfectly formed white eggs. These eggs are a true delight to collect and can be a charming addition to any breakfast table.

Silkie

The Silkie has fluffy feathers and a distinctive appearance. Originating from ancient China, the breed is also treasured for its calm disposition and charming personality. Its most distinguishing feature is its soft, fur-like feathers, which lack the typical hard structure found in other chicken breeds. Instead, the Silkie boasts silky plumage that is incredibly smooth to the touch, giving it an almost ethereal aura. Available in a variety of colors including white, black, blue, and buff, these chickens are truly a sight to behold.

Not only are Silkie chickens lovely, but they also make wonderful companions and are often sought after as pets and are especially popular among families and children. Their tolerant nature makes them suitable for handling, and they are often seen participating in poultry shows and exhibitions. Silkie chickens are known to form strong bonds with their owners and are even known to enjoy cuddling and being held.

In addition to their endearing personality traits, Silkie chickens are also known for their maternal instincts. They are renowned for their ability to brood and care for eggs, often taking on the role of foster mothers for other poultry breeds. This nurturing nature has earned them a reputation as excellent mothers, making them a valuable asset in breeding programs and for those looking to expand their flock.

While Silkie chickens are primarily kept for their ornamental qualities, they also provide a source of nutritious eggs. Though small in size, Silkie eggs are prized for their rich and flavorful yolks. However, it is worth noting that Silkie chickens are not prolific layers, and their focus tends to be on raising their young and establishing strong social bonds within the flock.

Sultans

Sultans, also known as the Serai Taook, is truly a regal and mesmerizing breed. Originating from the Ottoman Empire, these chickens were originally bred as ornamental birds for the Sultan's courts. With their striking appearance and elegant demeanor, it's no wonder they were considered a symbol of prestige and luxury.

One of the most distinctive features of the Sultan chicken is its lavish plumage. They have a profusion of soft, fluffy feathers that cascade down their entire body, resembling an elaborate feather boa. Their feathers come in a variety of colors, including pure white, black, blue, and even a striking combination of all three. The crest on their head is another prominent attribute, forming a majestic crown-like tuft of feathers.

Beyond their remarkable appearance, Sultan chickens have a gentle disposition. They are known for their calm and friendly nature, making them excellent companions for both novice and experienced poultry enthusiasts alike. Their amiable and non-aggressive behavior makes them a suitable choice for families and backyard flocks.

In terms of egg-laying capabilities, Sultan hens may not be the most prolific layers, but they do produce small to medium-sized eggs with creamy white shells. While their primary purpose is not egg production, their unique and captivating presence in the poultry yard more than compensates for any shortfall in egg numbers.

Due to their extravagant plumage, Sultans require a bit of extra care. Their feathers are delicate and prone to damage, so it's essential to provide them with a clean and safe environment. Regular grooming and monitoring of their feathers are necessary to ensure they remain in prime condition.

Chapter 3: Rare and Endangered Breeds

Barred Holland

The Barred Holland is also known as the Barred Plymouth Rock and with their striking black-and-white barred plumage can be difficult to distinguish from Plymouth Rocks and Dominiques.

In terms of their physical characteristics, Barred Holland chickens possess a medium-sized, well-rounded body with a broad and deep chest. Their feathers exhibit a unique barring pattern, which consists of alternating black and white stripes, giving them a distinct and attractive appearance. They have moderately sized single combs and bright red wattles and earlobes. Additionally, they possess yellow legs and feet, adding a vibrant splash of color to their overall appearance.

Barred Holland birds are known for their versatility and adaptability. They are highly capable of withstanding diverse climates and are known to be cold hardy, making them suitable for various geographical regions. These birds are also renowned for their excellent egg-laying abilities. They typically produce medium to large brown eggs, averaging around 200 to 280 eggs per year. This makes them a practical choice for those seeking a consistent supply of fresh eggs.

In terms of temperament, Barred Holland chickens tend to be calm, gentle, and sociable creatures that enjoy human interaction and are good around children, which makes them an ideal choice for families. With proper care and attention, they can become quite tame and even enjoy being held.

Buckeye

Developed in the late 19th century by a poultry farmer named Nettie Metcalf, the Buckeye breed is named after the state of Ohio, known as the Buckeye State.

One of the distinguishing features of the Buckeye is its stunning mahogany-colored feathers. With a lustrous reddish hue that shines in the sunlight, these feathers make the Buckeye stand out among other chicken breeds. Not only are they visually striking, but they also provide excellent camouflage against predators in the wild.

The Buckeye is also an American heritage breed, renowned for their hardiness and adaptability. They have a robust nature and are known to be quite self-sufficient, thriving in a variety of climates and landscapes. Whether it's scorching summers or freezing winters, the Buckeye chicken can withstand the elements with ease.

They are also known for their exceptional foraging abilities, eagerly exploring their surroundings in search of insects, worms, and vegetation. This natural inclination to forage contributes to their overall health and vitality, as well as reducing reliance on commercial feed.

Aside from their attractive physical traits, the Buckeye breed is also prized for its excellent meat production. Buckeyes are known to have flavorful and tender meat, making them a popular choice for farmers and homesteaders seeking a dual-purpose breed. Their meat is often described as juicy and full of flavor, making it a favorite for home-cooked meals.

Furthermore, Buckeyes have a docile and friendly temperament, making them great additions to a family-friendly flock. They are known for their calm and gentle nature, making them easy to handle and a joy to be around. Their sociable behavior makes them suitable for first-time chicken owners and those looking for a breed that interacts well with children.

As a heritage breed, the Buckeye chicken holds historical significance and plays a role in preserving our agricultural heritage. Despite facing near-extinction in the past, dedicated breeders and enthusiasts have worked tirelessly to revive and promote the Buckeye breed, ensuring its continued existence for future generations to enjoy.

The Buckeye ticks all the boxes.

Burmese

Originating from Myanmar (formerly Burma), one of the most distinguishing features of the Burmese chicken breed is their luxurious plumage. Their feathers display a stunning combination of colors, ranging from deep shades of chestnut and mahogany to rich gold and copper tones. This beautiful plumage, combined with their upright posture and graceful movements, makes them a true standout in any flock.

In addition to their visual appeal, Burmese chickens are cherished for their gentle and friendly character, making them wonderful additions to a backyard or homestead setting. Their pleasant temperament also makes them suitable for families with children, as they are generally tolerant and easy to handle.

Burmese chickens are also considered dual-purpose birds, meaning they are valued for both their egg-laying abilities and meat production. While they may not be the most prolific layers, their eggs are of high quality, with a rich and flavorful yolk. As for their meat, Burmese chickens are known to have a fine texture and excellent taste, making them a delectable choice for those seeking a unique culinary experience.

Health-wise, Burmese chickens are generally robust and hardy, able to adapt well to various climates and environments. With proper care and adequate living conditions, they can thrive and provide years of joy to their owners. Like any breed, they do benefit from regular health checks and a balanced diet to ensure their well-being.

Dong Tao

The Dong Tao chicken breed may be one of the most unique and fascinating poultry breeds you've ever come across. Also known as the Dragon Chicken, it is a fascinating and rare breed that originates from Vietnam. Known for its dramatic appearance and robust physique, the Dong Tao chicken is characterized by its incredibly large size and muscular build, making it a true marvel in the poultry world.

This breed is highly prized not only for its extraordinary appearance but also its delectable meat. The Dong Tao chicken is known for having incredibly thick, sturdy legs, which can sometimes be as wide as a human's wrist. These impressive legs, covered in scaly skin, are believed to have developed as a result of its long history of being raised for cockfighting. In addition to its distinctive legs and large, robust body its round head is adorned with bright red wattles and a vibrant comb. The breed's feathers come in various shades of gray, providing a striking contrast to its formidable stature.

While the Dong Tao chicken is primarily kept for ornamental purposes, its meat is highly sought after in Vietnam, known for its tender texture and rich flavor. Due to the breed's slow growth rate, the meat is considered a delicacy and is often reserved for special occasions and traditional feasts.

Due to its rarity and unique characteristics, the Dong Tao chicken is considered a symbol of prosperity and wealth in Vietnamese culture. It is often associated with good luck and is highly sought after by poultry enthusiasts and collectors worldwide.

However, the Dong Tao chicken breed faces challenges in terms of sustainability and conservation due to its limited population. Efforts are being made to preserve and protect this extraordinary breed, ensuring its survival for future generations to appreciate and enjoy.

Dorking

The Dorking breed stands as a true gem within the world of heritage poultry. Originating in the town of Dorking, Surrey, England, these chickens have been cherished for centuries, delighting enthusiasts with their distinctive appearance and exceptional qualities.

One of the defining features of the Dorking breed is its strikingly broad body, which is complemented by a stout and robust stature. With five toes instead of the usual four, including a notable fifth toe set far back on their feet.

These chickens also boast a beautiful variety of plumage colors, ranging from the elegant Silver Grey to the stunning Dark Cuckoo and the eye-catching White. Their feathers are dense and soft, adding to their appeal as an ornamental breed. Moreover, their small and neat rose comb, paired with their medium-sized wattles and earlobes, perfectly complements their overall aesthetic.

While the Dorking's visual qualities are impressive, their desirable attributes extend far beyond their appearance. Renowned for their excellent meat quality, these birds are often sought after by those who appreciate fine dining. With their tender and flavorful meat, Dorkings have gained a reputation as the ideal choice for gourmet cuisine.

Beyond their culinary acclaim, Dorkings also excel as egg layers. Though not as prolific as some modern breeds, they consistently produce a moderate amount of large and creamy white eggs, making them a practical and dependable choice for those seeking a sustainable egg supply.

Another remarkable characteristic of the Dorking breed is their calm and friendly temperament. Known for their docile and amiable nature, Dorkings are highly regarded for their ease of handling and their compatibility with both beginners and experienced poultry keepers alike. Their gentle disposition and willingness to engage with their human caretakers make them an excellent choice for families and individuals seeking a companionable chicken breed.

With their rich history dating back to Roman times, Dorking chickens have stood the test of time, maintaining their distinct attributes and captivating allure. Whether you are a poultry enthusiast, a heritage breed preservationist, or simply someone in search of a remarkable addition to your backyard flock, the Dorking chicken breed is sure to leave a lasting impression.

Houdan

Originating from the town of Houdan in France, these birds are known for their magnificent, crested head, adorned with a full, feathery crest that gives them an air of regal elegance. Their crests are often compared to the look of a fancy hat, making them a favorite among chicken enthusiasts and breeders.

Aside from their impressive crests, Houdans have an eye-catching appearance with their unusual, mottled feather pattern. The black and white speckled plumage, known as "millefleur," gives them a truly unique and beautiful appearance. This distinct feathering, combined with their crests, makes Houdans a favorite for exhibition and showmanship.

Houdans have admirable qualities as a productive breed as well. They are known for their excellent egg-laying capabilities, with hens consistently producing a good number of medium-sized white eggs throughout the year. Houdans are also known for their tender and flavorful meat, making them a versatile breed for both egg and meat production.

In terms of temperament, Houdans are generally friendly birds, making them suitable for backyard or small-scale farming. They are known to be relatively calm and gentle, allowing for easy handling and interaction. Their friendly nature also makes them a great choice for families with children or those seeking a breed that is easy to manage and handle.

Java

The Java breed, known for its elegant appearance and historical significance, originates from the island of Java in Indonesia with a heritage dating back to the 19th century. Java chickens are renowned for their graceful stature, with mature roosters weighing around 8 pounds and hens tipping the scales at approximately 6 pounds. These birds are known for their beautiful plumage, characterized by glossy black feathers that shimmer with iridescent greenish-purple undertones in the sunlight.

Java are celebrated for their versatility and are dual-purpose birds as well, excelling in both egg production and meat quality. Hens lay brown eggs with a rich flavor, and their consistent egg-laying capabilities make them a reliable choice for both farmers and backyard chicken keepers. The Java breed also possesses a well-developed muscular structure, resulting in flavorful and tender meat, making them a desirable option for those seeking a sustainable source of poultry.

In addition to their practical qualities, Java chickens possess a calm and friendly temperament, making them an excellent choice for families and hobbyists. They are known for their docility, which makes handling and care a breeze. Their adaptability to various climatic conditions further contributes to their appeal, as they can thrive in both cold and hot environments.

Preserving the Java chicken breed is of utmost importance to poultry conservationists, as they are considered rare and endangered. By raising Java chickens, enthusiasts not only benefit from their practical attributes but also contribute to the preservation of a breed with a significant historical legacy.

Scots Dumpy

The Scots Dumpy, also known as the Bakies, is a remarkably distinctive breed of chicken that originates from Scotland. True to its name, this breed stands out with its unique characteristic of having exceptionally short legs, giving it a noticeably squat and compact appearance. The Scots Dumpy's short legs, combined with its round body and full plumage, create a charming and endearing presence. With a history that dates back centuries, this breed has become a beloved icon in Scottish culture and a favorite among poultry enthusiasts worldwide.

Known for its friendly personality, the Scots Dumpy is an excellent choice for backyard flocks and small farms. It adapts well to various climates and is particularly suited to free-range environments. Despite its short stature, the Scots Dumpy is a sturdy and robust breed, capable of withstanding harsh weather conditions. Its low center of gravity makes it less prone to tipping over, providing an advantage in windy or uneven terrains.

The Scots Dumpy is not only visually appealing but also proves to be a practical choice for egg production. Although they may not lay as many eggs as some other breeds, the eggs are known for their excellent quality. The medium to large-sized eggs features a rich, creamy shell color, and the hens are known for being dedicated and attentive mothers if given the opportunity to brood.

Chapter 4: Dual-Purpose Breeds

Black Australorp

Originating from Australia, this breed has gained popularity worldwide, thanks to its exceptional egg-laying abilities and stunning appearance.

One of the most striking features of the Black Australorp is its beautiful, shiny black plumage. Its sleek feathers give it an elegant and distinctive look, making it a standout in any flock. Not only is this breed aesthetically pleasing, but it also boasts an array of desirable traits that make it an excellent choice for chicken enthusiasts and farmers alike.

Celebrated for its outstanding egg-laying capabilities, the Black Australorp holds the world record for the most eggs laid by a single hen in one year. These hens are prolific layers, consistently producing large, brown eggs with strong shells. With their dependable egg production, Black Australorps are highly valued for their ability to provide a plentiful supply of fresh, nutritious eggs to households and businesses.

In addition to their impressive egg-laying capacity, Black Australorps are known for a calm and friendly temperament. They are easy to handle and are often considered one of the most docile chicken breeds. This makes them an excellent choice for families with children or for those seeking a peaceful backyard flock. Their adaptable and sociable nature also allows them to integrate well with other chicken breeds, promoting harmony within a mixed flock.

Furthermore, the Black Australorp breed is known for its hardiness and resilience. Originally bred to withstand the harsh Australian climate, these chickens have developed a robust constitution and can thrive in various weather conditions. They are excellent foragers and possess strong immune systems, making them less susceptible to common poultry diseases.

Brahma

The Brahma is also known as the "King of Chickens," a magnificent breed that commands attention with its sheer size and unique appearance. Originating from the Brahmaputra region in India, Brahma chickens made their way to the United States in the mid-19th century, where they quickly gained popularity.

One of the most striking features of the Brahma chicken is its colossal size. These birds can grow to be one of the largest chicken breeds, with males reaching an impressive weight of about 12-15 pounds and females weighing around 9-12 pounds. Standing tall with strong, well-feathered legs, Brahma chickens exude an air of dignity and grandeur.

Their appearance is equally captivating. Brahma chickens have a bold and regal presence, with a wide, pronounced skull, small wattles, and a pea comb. Their eyes are set deeply and are often dark, adding to their intense and mysterious look. The feathers of a Brahma chicken are thick and profuse, coming in a variety of colors including light, dark, and buff. Some individuals even exhibit attractive feather patterns such as lacing or penciling.

Aside from their remarkable appearance, Brahma chickens also possess a delightful temperament. They are known for being gentle and friendly, making them a popular choice for families. Their calm and patient nature makes them great companions for children, and their tolerance towards other animals is remarkable.

Brahma are not only admired for their appearance and temperament but also valued for their egg-laying capabilities. While they may not be the most prolific layers, with an average of three to four eggs per week, their eggs are quite large and have beautiful brown shells. The rich, flavorful taste of Brahma eggs is a testament to their superior quality.

Delaware

The Delaware breed, originally developed in the United States, has beautiful white feathers and contrasting black neck and tail feathers, a visually stunning bird that instantly catches the eye. The breed's feathers are soft and glossy, giving it an elegant and regal appearance. It is not uncommon to see Delawares showcased in poultry exhibitions, where their unique markings and well-proportioned bodies make them a favorite among judges and spectators alike.

While their aesthetic appeal is undeniable, the Delaware breed also excels in egg production. Hens are known to lay a remarkable number of large, brown eggs throughout the year, making them a practical choice for those seeking a productive flock. Their eggs have a reputation for having excellent flavor and nutritional value, making them highly sought after by discerning egg enthusiasts.

In addition to their beauty and egg-laying capabilities, Delaware chickens are known to be quite calm and friendly and are generally easy to handle. Delawares also tend to be relatively low maintenance birds, adapting well to a variety of climates and environments.

Orpington

Developed in the late 1800s by William Cook in the village of Orpington, Kent, England, the Orpington breed quickly gained popularity for their exceptional beauty and gentle nature.

Orpingtons are known for their soft, fluffy feathers that come in various stunning colors including black, blue, buff, and white. With their round, plump bodies, and short legs, they have an adorable appearance that is hard to resist.

Not only are Orpingtons easy to look at, but they also make excellent pets and perfect for families with children. These chickens are also known for their ability to adapt well to different climates, making them a great choice for both urban and rural settings.

Whether you're looking for a lovely addition to your backyard flock or a charming companion, the Orpington chicken breed is sure to bring joy and beauty to your life.

Jersey Giant

Renowned for its impressive size and appearance, the Jersey Giant has earned its rightful place as one of the most captivating and fascinating breeds in the poultry world. Towering over other chicken breeds, the Jersey Giant is true to its name among its feathered counterparts. Originating in the United States in the late 19th century, this breed was developed to produce a dual-purpose chicken that could excel in both egg production and meat quality. Admired for its gentle demeanor and excellent foraging abilities, the Jersey Giant has become a favorite among backyard chicken enthusiasts.

Mature roosters can reach an astonishing weight of 13-15 pounds, while the hens are not far behind, weighing in at 10-12 pounds. This impressive stature makes them the largest purebred chicken breed in existence. Despite their large size, Jersey Giants are graceful and well-balanced, with a sleek and shiny black plumage or occasionally a stunning blue variation. The breed also boasts a deep, broad body, strong legs, and a neatly tucked tail, further adding to its imposing appearance.

While their size may be awe-inspiring, the Jersey Giant is not just a pretty face. These chickens are known for exceptional egg-laying capabilities as well, with hens typically laying rich brown eggs, also known for their large size, with some individuals even producing *jumbo-sized eggs*. With consistent care and proper nutrition, a Jersey Giant hen can lay up to 200 eggs per year. This makes them an excellent choice for those who appreciate a steady supply of fresh, large eggs.

Beyond their egg-laying prowess, the Jersey Giant is also valued for its well-flavored, tender meat. Thanks to their substantial size, these chickens yield a generous amount of flavorful, juicy meat, making them a preferred choice for those seeking homegrown, high-quality poultry for their table. Their excellent meat-to-bone ratio and tender texture have placed the Jersey Giant in a league of its own when it comes to meat production.

And, as if that wasn't enough, the Jersey Giant is also known as a good-natured bird as they are typically friendly and easygoing chickens that get along well with both humans and other flock members.

Sussex

Sussex chickens, originating in the Sussex region of England, are easily recognizable by their beautiful feather patterns which come in various color variations such as Light, Buff, Red, and Speckled. Their feathers are dense and provide excellent insulation, allowing them to thrive in different climates. With their medium-sized bodies and plump frames, Sussex chickens are also admired for their meat qualities.

Aside from their good-looks, Sussex chickens are highly regarded for their exceptional egg-laying capabilities. Hens can produce a remarkable number of large, brown eggs throughout the year, making them a reliable source of fresh eggs for households and egg-based businesses.

They are known to be gentle and friendly birds, making them suitable for families with children or individuals seeking a peaceful backyard flock. Their calm nature also makes them easier to handle and manage.

In terms of care, Sussex chickens are relatively low maintenance. They adapt well to different environments and are both cold and heat tolerant. With proper nutrition, housing, and routine care, these chickens thrive and exhibit good overall health.

Chapter 5: The Bantam Wonders

Known for its small stature and unique characteristics, Bantams have captivated chicken enthusiasts for centuries. Originating in Southeast Asia, these pint-sized chickens quickly gained popularity worldwide for their ornamental appeal and friendly demeanor.

Bantams come in a wide variety of breeds, each with its own distinct traits. From the feisty and colorful Old English Game Bantams to the adorable and fluffy Silkie Bantams, there is a Bantam breed to suit every chicken lover's preference.

1	Rosecomb Bantam	1.5 lbs
2	Silkie Bantam	4 lbs
3	Dutch Booted (Sablepoot) Bantam	2.2 lbs
4	Sebright Bantam	2 lbs
5	Japanese Bantam	1.5 – 2 lbs
6	Nankin Bantam	2 lbs
7	Buff Orpington Bantam	3 lbs
8	Barbu d'Anvers Bantam	1.5 lbs
9	Pekin Bantam (Cochin Bantam)	1.5 lbs
10	Barbu d'Uccle Bantam	1.5 – 2 lbs

One of the most endearing qualities of Bantam chickens is their petite size. They are typically about one-third to one-half the size of standard chicken breeds, making them perfect for those with limited space or urban dwellers. Their diminutive stature also adds to their charm, as they are often described as "miniature chickens" with personalities that match their size.

Despite their small size, Bantams are known for their remarkable personalities. They are often friendly, curious, and sociable, making them great additions to backyard flocks or as family pets. Bantam roosters are particularly famous for their bold and confident attitudes, proudly strutting around with their vibrant plumage.

Another interesting characteristic of Bantams is their ability to lay eggs. While they may not produce as many eggs as larger breeds, Bantam hens are known for their petite and colorful eggs, adding a touch of variety to any basket.

With their unique appearance, charming personalities, and ability to adapt to various living situations, Bantam chickens have become a favorite among chicken enthusiasts and hobbyists alike. Whether you're looking to add a beautiful ornamental breed to your flock or seeking a friendly and low-maintenance pet, Bantams are sure to bring joy and delight to your chicken-keeping experience.

About the Author

Alison Isaacs learned a lot from her maternal grandmother, Margaret and most of them were about how things were done in a time without running water and electricity. Although updated for the times, Alison brings her grandma's legacy to life with these simple how-to books. Good for folks who live in rural areas, those who want to live off-the-grid, and people who want to know how to better prepare when disaster strikes. Written in easy-to-understand language these self-help/how-to books are simple enough for a child to understand and yet profound enough for older generations.

www.ingramcontent.com/pod-product-compliance
Lightning Source LLC
Chambersburg PA
CBHW072229290526
45794CB00007B/2943